Copyright © 2020 by Eric Goldinger

All rights reserved. This book or any portion thereof may not be reproduced or used in any manner whatsoever without the express written permission of the publisher except for the use of brief quotations in a book review.

DISCLAIMER

The information in this book is not intended as medical advice. The content is for information and research purposes only.

By reading this disclaimer, you fully accept the terms of this disclaimer. If you are not in agreement with this disclaimer, please do not order or read this book. The content of this book is provided for information and educational purposes only. Kindly do not interpret this book or its content for a medication or product. This is only a book guide.

Thanks

Table of Contents

INTRODUCTION TO METHYLSULFONYLMETHANE 4
    Historical look to MSM 7
    Structure and Properties 10
    Role of Sulfur 11
    Types of MSM Supplements 13
    Top Food sources of MSM 14

MSM Amazing Benefits 16
    MSM for Joint Health 16
    Methylsulfonylmethane for Inflammation 24
    MSM for Neurological Diseases 26
    MSM for Bursitis 28
    MSM for Tendonitis and Tenosynovitis 34
    MSM for Acne 37
    MSM for Allergies 42
    MSM for Carpal Tunnel Syndrome 45
    MSM for Scar Tissue 47
    MSM for Indigestion 51
    MSM for Brittle Nails 54
    MSM for Hair Growth 55
    MSM for Snoring 59
    MSM for Hay Fever 62
    MSM and Cancer 64

MSM and Diabetes	71
MSM RECIPES	72
Where Is Your MSM Made?	75
Side Effects of MSM	77
WARNINGS	79
CONCLUSION	80

INTRODUCTION TO METHYLSULFONYLMETHANE

This mineral is magic! Dimethyl Sulfone a.k.a methylsulfonylmethane (MSM) is a chemical compound which has been steadily gaining in popularity over the last few years. And logically, more supplements containing this ingredient have started to emerge onto the scene. Providing a very basic and essential nutrient that most people lack, sulfur, getting more into your diet may be something to consider.

Many people are still unaware of dimethyl sulfone's benefits, or even existence for that matter. This basic supplement may be up there with things like magnesium, omega 3 fatty acids, Vitamin D, and Bone Broth as an everyday wellness strategy. MSM Sulfur is great for your overall health!

Methylsulfonylmethane (MSM) is also called dimethyl sulfone, DMSO2, methyl sulfone, and many other names. MSM is organic sulfur found in fruits, vegetables, grains, and milk as well as the urine of animals. It is an oxidation product of dimethyl sulfoxide (DMSO) and a source of sulfur for producing the amino acids methionine and cysteine.

People take MSM by mouth and apply it to the skin for chronic pain, osteoarthritis, joint inflammation, rheumatoid arthritis, osteoporosis, inflammation around the joints (bursitis), tendinitis, swelling around the tendons (tenosynovitis), nerve pain (neuropathy) caused by cancer drugs, musculoskeletal pain, muscle cramps, hardened skin condition called scleroderma, scar tissue, stretch marks, hair loss, wrinkles, protection against sun/wind burn, eye inflammation, oral hygiene, gum disease, wounds, cuts, and abrasions/accelerated wound healing.

People also take MSM by mouth for relief of allergies, chronic constipation, "sour stomach", ulcers, a bowel disease called diverticulosis, premenstrual syndrome (PMS), mood elevation, obesity, poor circulation, high blood pressure, and high cholesterol. It is also taken by mouth for type 2 diabetes, liver problems, Alzheimer's disease, to control snoring, lung disorders including emphysema and pneumonia, chronic fatigue syndrome, autoimmune disorders (systemic lupus erythematous), HIV infection and AIDS, and cancer (breast cancer and colon cancer).

MSM is also taken by mouth for eye inflammation, mucous membrane inflammation, temporomandibular joint (TMJ) problems, leg cramps, migraine, headaches, hangover, parasitic infections of the intestinal and urogenital tracts including Trichomonas vaginalis and Giardia, yeast infection, insect bites, radiation poisoning, and to boost the immune system.

Historical look to MSM

MSM itself was first gradually identified around the time of World War II, by researchers studying the contents of cow's milk. MSM is simply a particularly easy to absorb bioavailable form of sulfur, a geologically and biologically common mineral with central roles in many biological reactions. People have recognized sulfur for its healing qualities since ancient times, though we are today only barely beginning to understand how precisely sulfur contributes to our health.

Sulfur hot springs have been popular in both the old world and the new since long before recorded history, despite the rotting egg smell, and more than a few have been touted as a "fountains of youth" for their reported healing and age-defying qualities. Before the age of modern medicine, it was not uncommon for doctors to prescribe sulfur to patients with the instruction that it be poured into a hot bath to be soaked in. There is also a long tradition of using sulfur-rich plants, particularly, garlic, in ancient medicine. Cabbage, onions, horseradish, Brussels sprouts, leeks, mustard, and kale are all also foods that contain high numbers of organic sulfur compounds.

Though ancient humans long recognized sulfur for its healing qualities, it took modern science to rediscover and recognize the benefit of bioavailable sulfur. In the late 1950s, Dr. Stanley Jacob, head of transplantation research at the University of Oregon Medical School, encountered a sulfur compound known as dimethylsulfoxide-DMSO- while trying to find a way to freeze organic tissue without damaging it. While working with DMSO in the lab, he found that its medical use extended far beyond that of a cryoprotective agent. It showed analgesic and anti-inflammatory properties and promise as a skin-penetrating agent. Dr. Jacob hypothesized that DMSO would reduce pain when put over the location of an injury or localized bone or tissue illness.

He began testing it in his laboratory, and eventually started using it to successfully treat patients in his clinic who suffered from unresolved arthritis, bursitis, tendinitis, and other conditions. In 1963, he reported it to the meeting of the American College of Surgeons, and before long, DMSO was on the front page of The New York Times, touted as "the most exciting thing in medicine." By 1965, over 1,500 studies had been done involving about 100,000 patients, primarily regarding its usefulness in treating musculoskeletal injuries and diseases.

Unfortunately, due to a medical catastrophe involving Thalidomide, a widely used sedative drug that had been discovered to cause serious deformities to the fetus when taken during pregnancy, the FDA overhauled its procedures and postponed its consideration of many interventions.

One of the casualties of these changes was the approval of DMSO for medical uses, severely undercutting the source of the funding for all of the labs that had been researching it. It took five years to get DMSO approved for veterinary applications, where it became popular among veterinarians for treating a variety of musculoskeletal disorders in both large and small animals.

Structure and Properties

MSM is water soluble, the powder form is white and odorless, and the molecule is 34% sulfur by weight.

MSM is the oxidized byproduct of dimethylsulfoxide (DMSO), which is a process that occurs naturally in plants as the roots and foliage can absorbed DMSO; MSM may be degraded into the metabolite dimethylsulfide (DMS) but this is an instable and gaseous metabolite. DMS is actually the origin of MSM in plants as well, as microscopic plankton creates dimethyl sulfonium salts that convert to

DMS which is then delivered to plants via the air; photosynthesis and radiation converts DMS into DMSO.

$$CH_3-\overset{\overset{\displaystyle O}{\|}}{\underset{\underset{\displaystyle O}{\|}}{S}}-CH_3$$

Role of Sulfur

Sulfur is an abundant and fundamental nutrient in human cells. One of the most important functions of sulfur compounds and sulfur-containing amino acids is to assist in the production of glutathione. Glutathione is the body's most important antioxidant which regulates other antioxidants, detoxification, and helps to control inflammation. Glutathione also serves the vital role of protecting

DNA from becoming damaged, a potentially major source of aging and increased cancer risk.

Having inadequate levels of sulfur and glutathione, increase your susceptibility to chronic inflammation and genetic damage, potentially opening the doors for pretty much every other chronic disease people face today.

Some indications of low sulfur concentration include:

- Depression
- Fatigue
- Degeneration disease
- High sensitivity to physical and psychological stress

High dietary sulfur levels are thought to be fairly common in some of the healthiest cultures around the world. For example, Icelanders are known for being extremely healthy, having low rates of depression, obesity, diabetes, and heart disease. One potential explanation for this is the sulfur-rich volcanic ash that makes up much of the island. This sulfur content passes into drinking water, produce, and meat products.

Types of MSM Supplements

There are several types of MSM supplements available:

- Powder form
- Cream/lotion form
- Gel form
- Pill/tablet form
- Liquid eye drops
- Crystals

It's available in powder form most often. You might find it in powdered supplement mixes, such as green powders or digestive aids. MSM powder is believed to be the most absorbable kind.

Top Food sources of MSM

Several of the foods in your current diet contain MSM. Rich sources of MSM include:

- Coffee
- Beer
- Tea
- Raw milk
- Tomatoes
- Alfalfa sprouts
- Leafy green vegetables
- Apples
- Raspberries
- Whole grains

Other Sulfur rich foods

- Kale
- Cabbage
- Cauliflower
- Asparagus

- Mustard Green
- Onions
- Garlic
- Brussel Sprout
- Broccoli
- Bok Choy
- Nuts
- Radish
- Leeks
- Meat
- Seafood

MSM Amazing Benefits
MSM for Joint Health

Arthritis isn't actually a single disorder. It's an umbrella term that is used for informally referring to any one of over a hundred different diseases that cause pain in your joints. Some of these diseases are the result of mechanical damage to the joint itself, perhaps resulting from surgery or an accident. Some types of arthritis result from wear and tear on joints over a lifetime of use, and commonly strike aging athletes or people who have spent decades working with their hands for a living. While most types of arthritis begin in people once they approach middle age and grow gradually more severe over decades, some types of arthritis can strike with severity at a young age, usually as a result of a hereditary or biological abnormality, or due to extreme joint usage associated with some types of high impact sports.

Despite the great variety in causes of arthritis, most doctors use a one-size-fits-all approach to treating joint pain by prescribing pharmaceutical painkillers and NSAIDs. While these generally work to numb pain receptors in the short term or temporarily reduce inflammation, they do nothing for long term cartilage recovery. However, for most forms of arthritis, there may be other options.

The most common form of arthritis is osteoarthritis. This is a degenerative form of arthritis which results when cartilage between the bones wears away, causing the bones to painfully rub against each other. This causes pain, swelling, and stiffness. Cartilage is a firm, whitish connective tissue that cushions and lubricates between bones. Your outer ears are formed by elastic cartilage, but Synovial Joint the type of cartilage between your bones is actually more similar to the piece of firm, only slightly flexible cartilage that separates your two nostrils, known as hyaline cartilage. When hyaline cartilage separates two bones at a joint, it is referred to as articular cartilage. There are actually two pieces of articular cartilage in each joint, one attached to each bone. Between these two pieces of cartilage is a white, viscous substance called synovial fluid, which has a consistency similar to egg whites and plays a role not unlike bearing grease in a wheel, moving in and out of the joint as it flexes, all while lubricating the whole system. Finishing out the structure is the synovial membrane, which acts like an inner tube to seal the synovial fluid in place around the joint. The synovial fluid, in addition to its mechanical role in joint lubrication, also plays an important role in nourishing cartilage. Cartilage is unlike other tissues in the body in that it does not have its own blood supply. Instead, the synovial fluid delivers nutrients to the articular cartilage and keeps it clean and healthy.

Supplements that boost synovial fluid health also benefit cartilage protection and regeneration. One such supplement is methylsulfonylmethane, commonly known as MSM or "organic sulfur". Sulfur is a core chemical element necessary for biochemical functioning and an elemental macronutrient for all organisms. It is a vital building block of amino acids and the connective tissue in joints and cartilage. MSM also provides methyl groups, which support many essential metabolic processes in the body. Taking MSM boosts synovial fluid health, and also magnifies the efficacy of glucosamine when the two are taken together since it ensures that glucosamine is effectively transported into the articular cartilage. Glucosamine is a natural supplement that triggers the production of hyaluronic acid in synovial fluid and may actually help regrow damaged cartilage. Producing cartilage requires lots of sulfur, so MSM is also a crucial supplement for cartilage growth.

There are dozens of double-blind, placebo controlled scientific studies that demonstrate measurable improvements in the joints of people and animals who take MSM. This science backs up the claims of the thousands of people who have successfully used MSM to treat their debilitating arthritis or watched MSM breathe new life into their aging pets.

Unlike many of the drugs commonly prescribed for joint pain, MSM has no harmful side effects and will cause no lasting damage. The only unintended consequences you're likely to notice when you start taking MSM are healthier skin and nails and a stronger immune system, since MSM has shown significant benefits to these body systems as well.

Now that you know about the potential benefits of MSM, and about the lack of harmful side effects, you're likely eager to go ahead and give it a try. Before you go online and start comparing prices, however, you should be aware that with MSM, you get what you pay for.

The difference in quality between MSM distilled and packaged in the United States vs. MSM made overseas (frequently in China) is considerable. If you take a generic brand MSM, you risk inadvertently absorbing heavy metals and other pollutants, and could even do yourself more harm than good. Though many Chinese made MSM products will say that they are "100% pure," it's important, as with any supplement, to always do your research and make sure that what you're taking is what it says it is, and to buy from brands with a reputation for quality.

The types of pain which have been treated successfully with MSM include:

- Personal injury due to accidents, burns, etc.
- Osteoarthritis and rheumatoid arthritis
- Fibromyalgia

- Lower back pains
- Headaches, migraines
- Muscle aches
- Bursitis
- Tennis elbows and other sports injuries
- Carpal Syndrome
- Sclerosis
- Whiplash
- RSI (Repititive Strain Injury)
- Scars due to burns, operations, accidents, etc.

The way MSM impacts pain is currently explained by the following mechanisms:

- MSM is a natural analgetic: it blocks the transfer of pain impulses through nerve fibers (C-fibers).
- MSM blocks inflammations and inflammatory processes. MSM enhances the activity of cortisol, a natural anti-inflammatory hormone produced by the body.

- MSM improves the permeability of cell membranes. This improves the uptake of nutrients and many vitamins and the elimination of waste products and excess cellular fluids.
- MSM dilates bloood vessels, enhancing the blood circulation. This, too, helps to eliminate waste products from the body, which speeds up healing.
- MSM is a muscle relaxant. This is an important and often overlooked benefit of MSM. Many chronic pains are aggrevated by chronic muscle tension in the body.
- MSM aids the natural defense mechanisms in the body by regulating the prostaglandin metabolism, and regulates the formation of anitbodies and immune complexes.

Methylsulfonylmethane for Inflammation

Inflammation is your body's natural response to infection, and is characterized by redness and swelling in the affected area. Without inflammation wounds wouldn't heal, and infections would be a lot more dangerous. But too much of a good thing isn't always a good idea. Some diseases lead to excess inflammation, eventually causing chronic joint pain and stiffness (arthritis), along with other side effects.

It's believed that MSM inhibits NF-kB, a protein complex involved in inflammatory responses in your body. It also reduces the production of cytokines such as tumor necrosis factor alpha (TNF-α) and interleukin 6 (IL-6), which are signaling proteins linked to systemic inflammation

MSM reduces inflammation in the joints by inhibiting pro-inflammatory mechanisms in the body. Glutathione, a key antioxidant, relies on sulfur to work properly in the body. MSM plays a critical role in detoxifying the body by regulating the production of glutathione.

MSM has therapeutic value in treatment of ulcerative colitis. It also have anti-inflammatory effects against lipopolysaccharide-induced inflammatory responses in murine macrophages.

Lipopolysaccharide toxicity or endotoxin are responsible for various health problems including obesity, autoimmune diseases, has negative impact on kidney, adrenal glands and lungs, damage to the endothelial layer of blood vessels that reduces the cardiac function and can lead to septic shock.

MSM is a powerful anti-inflammatory due to its ability to allow metabolic wastes to be removed from the cells. Excess weight on the body is actually inflammation. The cells of the body are chronically inflamed and retaining the byproducts of metabolic processes. Sulfur needs to be present In order for these toxins and wastes to be removed from the body. When these byproducts can be removed from the system, then the cells can also dispose of excess fluids that were being stored as a buffer. What results is a natural and effective reduction in unwanted weight.

MSM for Neurological Diseases

The brain is extremely sensitive to the effects of toxic materials such as heavy metals and organic compounds. Many of these compounds tend to accumulate in nerve cells where they can cause severe oxidative damage. Neurological disturbances may be the result. MSM is one of the few antioxidants which can easily pass the blood-brain barrier. It prevents and repairs oxidative damage and restores cell membrane elasticity and permeability. This allows the nerve cells to start excreting waste products.

The powerful action of MSM is illustrated in the following example. An older woman suffered from poisoning caused by exposure to aluminum. This heavy metal had accumulated in her brain and caused severe neurological damage. The woman had been confined to her bed for six years, unable to communicate with her surroundings. All this time she had not spoken a word. Medical doctors could no longer help her and had given up on her. She was completely dependent on her husband who took complete care of her needs. A natural health care practitioner advised two teaspoons of MSM (about 15 grams) daily. The MSM passed the blood brain barrier, and restored the permeability of the brain cell membranes, allowing her brain cells to purge the heavy metal poisons. Two weeks later, the orthomolecular physician prescribed a warm bath to help her eliminate the released poisons through the skin. After twenty minutes in this bath, the woman suddenly smiled and said

"Gee, I feel much better now." These were the first words she had spoken in years. Several months later, the woman was capable of leading a normal life again

MSM for Bursitis

Bursitis means inflammation of the bursa. A bursa is a sac-like structure that contains a lubricating fluid. A bursa is located anywhere you need a lubricating cushion-like where a muscle or tendon rubs over a bone or another muscle.

For example, a person who throws a ball too many times at one interval may begin to cause this friction process in the shoulder or elbow. Another way that bursitis occurs is when compression of the bursa happens on a regular basis, such as a side sleeper who places a lot of shoulder or hip pressure when they lay down for long periods. In this case it would not take too much movement of this already compressed area to cause friction of the bursa.

Normally a bursa does its job unnoticed, but if you engage in some strenuous activity, for example, it can let you know exactly where it is. When a bursa is repeatedly irritated, the body begins to deposit calcium spicules in that location (often these deposits can be seen on X-rays). The spicules are like ground glass in the bursa, and the more you move that part of your body, the more intense the pain.

Types of Bursitis

Even though hundreds of bursa exist throughout your body, there are some typical problem areas. Here are the five common types of bursitis and the symptoms associated with each.

- Sub-deltoid bursitis. One of the unfortunate byproducts of dusting every piece of furniture in the house, washing and waxing your car, cleaning windows, working long hours in the garden, etc., is an inflamed shoulder bursa. When you have sub-deltoid bursitis, sleeping on your side can be difficult and just moving your arm can be extremely painful.

- Trochanteric bursitis. Another common and painful type of bursitis involves the bursa over the hip bone. If you move your hand along the side of your leg, you can feel a bony knob that sticks out from the hip bone (the trochanter of the femur bone). The muscles and skin that pass over this knob are lubricated by this bursa. Among other things, a long walk or exercise class after a sedentary winter can inflame this bursa. Trochanteric bursitis often makes walking difficult and painful and sleeping on your side impossible. It's also worth noting that the pain from this area is sometimes confused with sciatic nerve pain.

- Olecrenon bursitis. Your elbow can be another site for bursitis problems. Though not as common as sub-deltoid and trochanteric bursitis, it can leave you looking like you have a golf ball under the skin.

- Pre-patellar bursitis. Your knee is loaded with bursa. This type of bursitis is often referred to as "housemaid's knee," since constant kneeling from scrubbing floors can cause it. Although most people mop their floors nowadays, making true "housemaid's knee" pretty obsolete, pre-patellar bursitis is still common among carpet layers and gardeners who typically spend long hours on their knees.

- Calcaneal bursitis. High heel shoes can claim credit for many cases of this type of bursitis, which affects the bursa in the heels. Shoes that are too tight or too large can also cause excessive pressure or rubbing on this bursa.

With proper care for the area, the pain in the bursa should lessen over three weeks, but it should be noted that the healing of the area continues and doesn't even peak until at least six weeks following the initial injury. This is due to scar tissue formation, which initially acts like the glue to bond the tissue back together. Scar tissue will continue to form past six weeks in some cases and as long as a year in severe cases. After 6 months this condition is considered chronic and much more difficult to treat. The initial approach to treating a bursitis is to support and protect the bursa by bracing any areas of the tendon that are being pulled on during use, as this will help stop bursa friction from occurring. It is important to loosen up the tendons, lessen the pain, minimize any bursa inflammation, and reduce the compression that can occur with lying down or sitting.

The pressure can be reduced while sleeping by using a softer bed topper like a memory foam mattress pad or even getting a new mattress that is composed with memory foam and/or latex foam. Memory foam and latex foam reduce compression because they are the only substances that conform to the bumps and curves of the body and can thereby reduce the pressure spot by more evenly disbursing the weight of the body.

Reducing bursa inflammation and soothing the pain of bursitis can be done topically if a pain reliever has the ability to penetrate the skin barrier and contains anti-inflammatory agents. A topical formula that contains natural menthol will not only relieve pain, but dilate the blood vessels. This allows for relief of the bursitis, without causing any stiffening of the tissue. MSM, also known as Methyl Sulfonyl Methane is a natural supplement that is getting a lot of attention due to its role in tissue healing at a cellular level. It is natural organic sulfur that comes from rain fall and is found naturally in the human body. It has been shown to have anti-inflammatory effects on the tissues.

Prevention of this condition requires routine stretching the affected joint, as well as any tightened muscles above and below on a regular basis and thereby lengthening the tendon connections around the bursa. This will allow less friction to the tendon/bursa/bone connection. As mentioned previously, reducing the compression with sitting or lying down is crucial for helping to keep the bursa from losing its natural lubrication and allowing the bursa friction to occur more readily. When bursitis does occur, it is important to treat it immediately, and thereby prevent it from reaching a stage that is more severe.

MSM for Tendonitis and Tenosynovitis

Inflammation of a tendon (tendonitis) and the lining of the sheath (tenosynovitis) usually occurring simultaneously. The synovial-lined tendon sheath usually is the site of maximum inflammation, but the inflammatory response may involve the enclosed tendon (e.g. as a result of calcium deposit).

A tendon is a tough and flexible band of fibrous tissue. The tendon is the part in your body that connects your muscle to the bones. The skeletal muscles in your body are responsible for moving your bones. When a muscle contracts it pulls on a bone to cause movements. The structure that transmits the force of the muscle contraction to the bone is called a tendon.

Tendons come in a variety of shapes and sizes. The very small are like the ones that cause movements in your fingers, other larger tendons like the ones controlling movement in your arms and legs. When functioning properly, these tendons glide easily and smoothly with muscle activity. Inflammation of the sheath can cause an entrapment of the tendon, as in the case of trigger finger.

Tendonitis also spelled tendonitis is the inflammation of the tendons and the muscle surrounding the tendon. Tendonitis can also be associated with inflammatory diseases that occur throughout your body, such as rheumatoid arthritis. Tendonitis can cause significant pain and is the most common injury in athletes over the age of 40.

Methylsulfonylmethane (MSM) stimulates the immune response, reduces inflammatory reactions and stimulates the healing of connectives tissues such as tendons and ligaments. MSM also has a positive effect on the repair and suppleness of the muscles.

MSM for Acne

Acne is a skin condition that occurs when your hair follicles become plugged with oil and dead skin cells. It often causes whiteheads, blackheads or pimples, and usually appears on the face, forehead, chest, upper back and shoulders. Acne is most common among teenagers, though it affects people of all ages.

If you've ever suffered with acne, you know all too well that it is a really hard ailment to fix. You've tried it all...from baking soda to Proactive. In the beginning, you notice improvement and then all of a sudden, it's not working anymore. It's almost like you've grown immune to it. Or it is causing you to break out even more and then you hear, "oh, it's gotta bring the bumps out first". What? I always wondered was that true. The jury is still out on that.

One of the most consistent acne remedies that I've come across is MSM (or methylsulfonylmethane).

Due to sulphur being involved in so many different processes in your body, increasing your sulphur intake has a number of health benefits. As a result, MSM supplementation can actually improve acne symptoms via three different pathways.

MSM plays a major part in the body's ability to produce collagen. As an essential ingredient in tissue to hold the human body together, collagen provides structural support and impacts the flexibility and resilience of skin. The human body produces collagen naturally, but this production tends to decline in adulthood, starting in the 20s and 30s. That's why skin naturally starts to sag and wrinkle the older we get. But with a collagen boost from regular MSM use, you could very well retain your youthful resilience.

Because MSM helps reinvigorate collagen production after it starts to decline, it's an excellent ingredient to include in any skincare routine to fight the signs of aging right at the source. Using MSM from a young age may even help reduce future signs of aging before they start.

As a suitable treatment for a variety of skin conditions that cause reddened, swollen skin, MSM can be applied to the skin to reduce puffiness, discoloration and swelling. Most acne treatments address clogged pores with harsh chemicals, but MSM offers a gentler way to treat inflammation associated with acne.

Besides reducing the signs of aging, collagen is helpful in the treatment of acne. Increased collagen in the skin helps repair acne-damaged skin tissue that could result in facial scarring.

When used in conjunction with proper cleansing, gentle exfoliation and regular moisturizing, topical MSM can have a dramatic, positive effect on the smoothness of the skin. Smoother, tighter skin with less puffiness and fewer wrinkles make skin appear younger without the need for potentially dangerous, invasive treatments.

Whether you're fighting wrinkles or acne or you simply want to take excellent care of your skin, you can reap the benefits of MSM's ability to reduce swelling and roughness while stimulating collagen-driven skin repair.

How to Use MSM For Acne

You can purchase creams or even make your own MSM paste and apply it to your skin topically, which may be a good idea in some cases when more targeted relief is needed. However, topical application is unlikely to provide all of the benefits discussed above.

Oral ingestion of MSM in the form of a powder or tablets is the simplest and likely to be the most effective method for improving your acne. Dosage ranging from 2-6g daily has been demonstrated to be safe for most individuals.

However, it is recommended to take a cautious approach and begin on the low end of that range. MSM is a natural compound so most healthy people will not need to worry about side effects but it is best to assess your tolerance first and adjust your dosage based on the results.

Individuals who are currently taking blood-thinning medication are advised against taking MSM supplements. To be totally safe, it is always recommended that you discuss any new supplementation or dietary changes with a registered health care professional first.

MSM for Allergies

Allergies are the sixth leading chronic condition in the U.S. and over 50 million Americans have them. If you have seasonal allergies and over-the-counter medications aren't helping, there may be a better alternative. Ranging from Sneezing, congestion, and migraines; Itchy throat, eyes, mouth, and skin; Eczema, rosacea, and other rashes; Sensitivity to chemicals, perfumes, pets, dust, and mold. Allergies strike in many forms. Allergies vary from mild to life-altering and may be seasonal or year-round. What they all have in common is that at their most basic, allergies are caused by hypersensitive histamine reactions to something in the environment that would cause little or no problem for most people.

MSM has demonstrated anti-allergic properties for decades; a small but significant segment of the MSM consumer base has used MSM primarily for treating allergies. However, it's only been relatively recently that scientists have begun understanding why MSM works in this way.

Sulfur compounds facilitate dozens of bodily functions. MSM is the most prolific sulfur compound in the human body and is crucial to many immune responses. Introducing more MSM into your systems may alleviate allergies in several ways, including:

- Methylsulfonylmethane inhibits histamine receptors inside your nasal passages. It blocks excess histamine absorption, which causes allergic reactions.
- Sulfur compounds promote better blood flow and reduce fluid buildup, which causes nasal inflammation. These compounds may also relieve allergy-related pain.

- MSM facilitates cell detoxification and reproduction. You may also feel energized after taking MSM supplements.

Sulfur compounds have a positive effect on your immune system. Their role as an anti-inflammatory can help you recover more quickly from a workout and MSM reduces oxidative stress that makes you age faster. This beneficial feature reduces inflammatory compounds like IL-6 and TNF-α to bolster your immune system.

These anti-inflammatory properties also help you improve your sinus condition by alleviating the following allergy symptoms:

- Headaches
- Runny nose
- Watery eyes
- Sneezing
- Nasal pain
- Nasal inflammation

- Congestion
- Fatigue

There's a wide range of recommended daily dosages, varying from 2,000-6,000mg. Age, body size and the level of MSM in your blood are contributing factors, and you can start at the lowest dosage and add more as needed. Many people notice changes within three weeks. When you use MSM, the sulfur compounds soften cell walls, letting allergens and free radicals to flow out of your system.

MSM for Carpal Tunnel Syndrome

Carpal tunnel syndrome (CTS) is an injury caused by a pinched nerve in the wrist. The injury causes pain and numbness in the index and middle fingers and weakness of the thumb. Carpal tunnel receives its name from the eight bones in the wrist, called carpals, which form a "tunnel" through which the nerve leading to the hand extends.

Signs and symptoms of CTS include:

- Nighttime painful tingling in one or both hands, frequently causing sleep disturbance.
- Feeling of uselessness in the fingers.
- A sense that fingers are swollen even though little or no swelling is apparent.
- Daytime tingling in the hands followed by a decreased ability to squeeze things.
- Loss of strength in the muscle at the base of the thumb near the palm.
- Pain shooting from the hand up the arm as far as the shoulder.

There are a number of ways to treat this syndrome, one method which has had great success is MSM (Organic Sulphur). Using MSM both orally and applying topically (on your body) will give the best chance at relieving strong pain. This equates to taking one teaspoon a day (This is an average dose and can be increased if necessary to accommodate pain levels)

MSM for Scar Tissue

Scar tissue refers to thick, fibrous tissues that take the place of healthy ones that have been damaged. Healthy tissues may be destroyed from a cut, significant injury, or surgery. Tissue damage may be internal, so scar tissue can form post-surgery or as a result of disease.

In the early stages, scar tissue isn't always painful. This is because nerves in the area may have been destroyed along with healthy body tissues.

But over time, scar tissue may become painful as nerve endings regenerate. Scar tissue can also become painful over the course of an internal disease. The amount of pain can also vary based on the severity of the initial wound as well as its location on your body.

Sometimes scar tissue can be painless. When it comes to scar tissue on your skin, you may notice it has a thicker texture compared to the rest of your body and that's it.

On the other hand, exterior scar tissue can be painful. Some of the symptoms of scar tissue pain include:

- Inflammation (swelling)
- Redness
- Itchiness
- Throbbing
- Sensitivity (to the touch)
- Reduced range of motion
- "creaky" sounds or sensations

Scar tissue you can't see may form due to internal wounds, surgeries, or underlying diseases. You may still feel pain and stiffness at these sites, especially if the scar tissue starts affecting the surrounding joints. Such is the case with knee or spinal scar tissue, as well as scar tissue formed following surgeries of the face, or from medical procedures like hysterectomies.

MSM is a wonderful supplement with a low toxicity rating that naturally occurs in the body and targets a host of issues. Some of the problems it may help with include hair, skin, nails, muscle fatigue, allergies and scar tissue trouble. I have also heard of people with cancer utilizing the supplement. Once upon a time, there was a big craze about DSMO and how it helped arthritis sufferers. It has been observed that the solvent DSMO converts to MSM when it hits the blood stream. Taking MSM may be equivalent to using DSMO but without the side effects and problems that come with utilizing an industrial grade solvent.

Some of the patents I read on MSM suggest that MSM basically supplies the body with a usable form of sulfur, which can be used to repair connective tissues, aid in joint problems and potentially energize a system that pulls allergens out of our bodies. Studies have been conducted on MSM that seem to indicate that there has been a benefit to animals with joint problems.

If you have significant scarring, MSM may help to reduce the visible scarring. It may help to give the skin the sulfur it needs to help in restructuring; it may also help in preventing scars. So if you have a burn or jagged cut, it might be helpful to take MSM during the healing process.

The sulfur is useful in the process that is used to build skin; one theory is that if there isn't enough sulfur available the process is modified resulting in scar tissue. Hence, if you have enough sulfur, you could avoid scarring as tissue regrows, according to that theory.

MSM for Indigestion

Indigestion also called dyspepsia or an upset stomach is a general term that describes discomfort in your upper abdomen. Indigestion is not a disease, but rather some symptoms you experience, including abdominal pain and a feeling of fullness soon after you start eating. Although indigestion is common, each person may experience indigestion in a slightly different way. Symptoms of indigestion may be felt occasionally or as often as daily.

Indigestion can be a symptom of another digestive disease. Indigestion that isn't caused by an underlying disease may be eased with lifestyle changes and medication.

People with indigestion may have one or more of the following symptoms:

- Early fullness during a meal. You haven't eaten much of your meal, but you already feel full and may not be able to finish eating.
- Uncomfortable fullness after a meal. Fullness lasts longer than it should.
- Discomfort in the upper abdomen. You feel a mild to severe pain in the area between the bottom of your breastbone and your navel.
- Burning in the upper abdomen. You feel an uncomfortable heat or burning sensation between the bottom of your breastbone and your navel.
- Bloating in the upper abdomen. You feel an uncomfortable sensation of tightness due to a buildup of gas.
- Nausea. You feel as though you want to vomit.

Less frequent symptoms include vomiting and belching.

MSM can help rebuild the lining of the digestive tract and lower inflammatory responses in response to allergic reactions to certain foods.

It's also useful for helping treat leaky gut syndrome since it can help stop particles from leaching out the gut through small junction openings, where they can enter the bloodstream and ignite an inflammatory response. This is due in part to the sulfur in an MSM supplement, which is important for digestion.

Additionally, MSM supplements seem to help treat hemorrhoids, according to studies. Applying a gel containing MSM along with tea tree oil can reduce pain and swelling caused by hemorrhoids (swollen blood vessels of the rectum that make it hard to go to the bathroom without pain or bleeding).

MSM for Brittle Nails

Many medical conditions can affect the shape or texture of the fingernails. Brittleness of the nails, meaning that the nails easily become cracked, chipped, split, or peeled, can be observed as a sign of aging or in response to the long-term use of nail polish or exposure to moist conditions (including frequent swimming or dishwashing). Some diseases are also associated with changes in the nails, which can include brittleness. Thin and brittle nails can be a sign of hypothyroidism, for example. The term onychoschizia refers to splitting of the fingernails as well as brittle or soft nails.

MSM provides fast benefits for nails too, as nail thickness, clarity and growth factors are dependant on the availability of Keratin. Again, when low in MSM / Organic Sulfur, the body struggles to produce enough Ketartin, which can result in weak brittle nails.

By simply increasing the oral intake of MSM, many see rapid impacts to nail growth & strength. MSM/Organic Sulfur crystals can also be used to create easy nail soaks that can have rapid and significant impact on nail health.

MSM for Hair Growth

MSM (also known as Organic Sulfur), is an excellent, natural alternative for increasing hair growth, and overall hair health, thickness and appearance.

MSM is also an amazing overall beauty support that helps nails grow thick and strong, and is outstanding for skin, complexion, wrinkles and acne.

MSM is vital for thick, lush hair, and can help with hair regrowth. It's also crucial for overall appearance and health of the skin, and nails. For anyone looking to maintain appearance naturally and effectively, MSM is what you seek.

The hair growth cycle consists of three stages:

- Anagen (growing),
- Telogen (intermediate), and
- Catagen (resting or shedding).

As we age, the Anagen stage gets shorter and shorter, which is why the older we get the more hair we lose. MSM has been shown in studies to help extend the body's natural hair growing cycle the 'Anagen' phase.

To compound hair loss and thinning hair, when MSM is deficient in the body, hair loss and thinning can accelerate. Many do not realize that MSM/Organic Sulfur should actually be the third most abundant mineral in the body behind Calcium and Phosphorus. That fact alone should tell you how important Sulfur actually is to the body and appearance, and how much the body actually depends on Sulfur.

Studies have shown that the majority of people are actually very low in MSM (Sulfur), due to the low amounts now found in food due to our farming and food processing methods today. Organic Sulfur/MSM is what is called an 'essential mineral', meaning you can't make it, you have to get it from either food or supplementation. Most people should be supplementing MSM daily for its vast benefits for appearance, health and general wellness.

MSM helps increase hair growth by aiding in the suppression of the Telogen (resting) phase, and helping to reactivate the Anagen (hair growth) stage.

This is thought to happen when people increase intake of MSM, as added Sulfur starts to build up in the middle layers of the scalp and follicle, which act as a bio- signal for the body to restart the Anagen cycle.

MSM/Organic Sulfur is also the primary building block of the protein known as "Keratin" which is responsible for the health and appearance of hair (and nails). Keratin is a fibrous, sulfur-based protein that provides the building blocks and structure for hair, and nails. Keratin is fully dependent on adequate MSM/Organic Sulfur in the body. With so many deficient in MSM/Sulfur, there can be a significant lack of Keratin in the body leading to many issues with thinning hair, shedding, as well as nail and skin problems.

Organic Sulfur (MSM) also helps the body create flexible "disulfide bonds" which act as the 'glue' that gives Keratin (and Collagen) strength and flexibility yielding stronger, soft more lush hair. Synthetic hair treatments & medications simply don't get to the "root" cause (Pun not intended).

Many women are led to believe these types of expensive products will help because of big budget advertising campaigns. These brands focus only on synthetic concoctions, external keratin serums, creams and leave-in treatments, because they can't patent naturally occurring minerals and nutrients like MSM.

As a result, these brands are not focused on 'natural solutions' that work by giving the body what it needs to fix the issue, and provide typically excellent results for many women and men. Instead, many mainstream hair and skin products are actually filled with serious toxins and harsh ingredients that do a lot more damage than good.

MSM for Snoring

Snoring. We've all done it at some point in our lives. If it is happening regularly each night, and causing disruption to other people's sleep, then it might be worth looking into why it is happening.

Snoring is the hoarse or harsh sound that occurs when air flows past relaxed tissues in your throat, causing the tissues to vibrate as you breathe. Nearly everyone snores now and then, but for some people it can be a chronic problem. Sometimes it may also indicate a serious health condition. In addition, snoring can be a nuisance to your partner.

Lifestyle changes, such as losing weight, avoiding alcohol close to bedtime or sleeping on your side, can help stop snoring. Common causes of snoring, some of which include;

- Allergies
- Eating too much at night
- Nasal Congestion
- Being overweight
- Sleep position
- Swollen adenoids or tonsils, especially in children

There are quite a few options in regards to treatment for snoring, a great one to consider is visiting a Sleep Physician who specialises in sleep disorders and can make an informed recommendation and suggested treatment.

In addition to this, there are also some lifestyle changes that can also help. Considering a weight loss program if this is a factor may help to decrease snoring as will stopping smoking. Not only does smoking pose big health risks, but "tobacco smoke causes the walls of airways to retain fluid and swell (this is called oedema). This causes the airway to narrow, worsening snoring. Smokers are 4 to 5 times more likely than non-smokers to suffer from snoring and obstructive sleep apnoea.

Incorporating nasal treatments is also another option. In particular, a nasal spray containing MSM (source of organic sulphur) could assist. In a trial using MSM, sufferers of chronic snoring where given 15% MSM drops in a water solution in each nostril 15 minutes prior to bed. 80% reported reductions in snoring with no side effects whatsoever in 90 days. To make up a solution add 15 grams or 3 level teaspoons of MSM to 85mls of boiled, filtered water. Spray up each nostril around 10 minutes before sleep each night.

If you're waking up tired and not feeling you're getting the restful sleep you need then perhaps snoring might be the culprit and a further look into why this might be happening could help you find ways to reduce and prevent it.

MSM for Hay Fever

Hay fever is a common allergic reaction which occurs at particular times of the year. It is known as seasonal rhinitis, sharing symptoms with perennial (year round) allergic rhinitis, but occurring as a reaction to pollen from grass, trees and weeds during the early spring and summer months. It can affect both adults and children.

It is caused when the body makes allergic antibodies (IgE) to certain substances, such as pollen, house dust mites or mould, which are known as allergens.

Grass pollen is the most common allergen (May to July), but tree (February to June) and weed (June to September) pollens can also cause the allergic reaction we know as hay fever. In perennial allergic rhinitis the symptoms continue all year round and usually relate to indoor allergens, such as house dust mites, pets, including birds, or moulds. Symptoms include:

- Itchy eyes/ throat
- Sneezing, blocked/runny nose
- Watering, red eyes (allergic conjunctivitis)
- Headaches, blocked sinuses
- Shortness of breath
- Tiredness
- The sensation of mucus running down the back of the throat, which can also be a symptom, is called 'post-nasal drip'.

"MSM has so many benefits for allergy sufferers that it's hard to know where to start. In one study, 55 volunteers diagnosed with seasonal allergies were given 1,300mg of MSM twice daily for 30 days. A significant reduction in symptoms was seen.

"As long as you're still suffering from any allergic symptoms, or are in pain, it's well worth supplementing MSM on a daily basis.

MSM and Cancer

Organic sulfur IS MSM, but MSM is not necessarily organic sulfur. Do not buy MSM unless you are certain it is pure organic sulfur (i.e. it has no fillers or chemicals).

"Organic sulfur water" is made by putting 16 TABLE spoons of true organic sulfur into a gallon jug of purified water (which has ZERO chlorine in it) or distilled water. If you use distilled water it can actually help detoxification, such as helping to remove heavy metals. If you use tap water, either use the tap water from the "hot" faucet (and let it cool off) or let the water sit for at least an hour before putting any organic sulfur in it.

NOTE: MSM is almost as safe to drink as water (based on LD-50). An adult could double, triple or even quadruple the doses of "MSM water" in this article and still be perfectly safe. Doses for taking the "MSM water" (taken from a gallon jug of water with MSM put into it) are discussed below.

For those who have had chemotherapy, or are currently on chemotherapy, they need to take "MSM water" every day to help the chemotherapy target the cancer cells. This will also be explained below. True organic sulfur can help deal with pain (because it gets oxygen inside the cells) and it can help deal with inflammation for the same reason. But this oxygen also can kill the microbes inside the cancer cells, which in turn can revert the cancer cells into normal cells.

Organic sulfur can also take the "trash" out of the cell every 12 hours. It helps get rid of heavy metals, which is important for many cancer patients and most patients with brain disorders. This also means that cancer patients with brain fog may benefit from this treatment, depending on what is causing the brain fog. While organic sulfur is a superb cancer treatment by itself, adding Vitamin C is like adding chocolate sauce to vanilla ice cream.

Knowing the difference between true "organic sulfur" and inferior MSM is very important. Air is 21 percent oxygen. Water is 89 percent oxygen. The oxygen from air gets to the lungs, heart, and muscles. However, it is the oxygen from water that needs to get inside the cells to make the cells healthy, give them energy and to kill microbes inside the cells.

True organic sulfur is what grabs the oxygen from the water and transports the oxygen inside the cells. This produces a surge of oxygen into the cancer cells. There is no way to describe the importance of true organic sulfur in the treatment of cancer, brain disorders, and many other health conditions. Every part of the human body (including the brain) is made of cells. Getting sulfur and oxygen inside these cells is critical for all aspects of health.

As far as cancer is concerned, it is used in the following ways:

- As an oxygen transport,
- To kill microbes in the bloodstream,
- To help sweep lactic acid out of the bloodstream to prevent cachexia,
- To get microbe-killing substances (such as Vitamin C) inside of cancer cells to help revert cancer cells into normal cells,
- To reduce swelling and inflammation (e.g. for brain cancer) and

- To help reduce pain (via removing lactic acid from the blood).

Dealing with lactic acid (which is frequently a major cause of pain) and the cachexia cycle, and dealing with the microbes in the bloodstream should be a TOP priority after a patient is stable (and quite frankly, from day one after diagnoses of cancer). The "lactic acid cycle" (i.e. cachexia) is caused by the following:

- Glucose enters into the cancer cells and is converted into lactic acid and is released into the bloodstream,
- The lactic acid (via the bloodstream) goes to the liver and the liver converts it into glucose,
- The glucose goes back to the cancer cells and the cycle starts again (go to Step 1).

Both end points of this cycle consume enormous amounts of energy.

There are several ways to deal with the deadly cachexia cycle:

- MSM removes lactic acid in the bloodstream, thus the lactic acid cannot get to the liver,
- Hydrazine sulfate blocks the conversion of lactic acid to glucose in the liver,
- Anything that kills cancer cells helps block the cycle (fewer cancer cells means less lactic acid),
- Anything that reverts cancer cells into normal cells also helps block the cycle (ditto).

In addition, the lactic acid cycle makes cancer patients weak, causes massive pain and can, therefore, reduce their will to live.

Lactic acid can be just as dangerous as the cancer cells.

While hydrazine sulfate does an excellent job of dealing with stopping the production of lactic acid, few people like to use it because of its numerous safety warnings, meaning it almost has to be used by itself.

MSM by itself, at high doses (two tablespoons of the crystals), is a superb cancer treatment for advanced cancer patients because of its ability to get rid of lactic acid in the bloodstream, kill microbes in the bloodstream, etc. Several experiments have shown that oral administration of MSM can protect rats against the onset of cancer. In one study, rats specially bred to be susceptible to breast cancer when given certain carcinogenic compounds were fed a diet containing MSM for a period of eight days. The control group did not receive MSM. Following this preliminary period, all rats were given oral doses of cancer-causing agents. There was no statistical difference in the number of tumors developing in the two groups. However, the MSM diet rats developed their first tumors some 100 days later than the control rats, and these tumors became cancerous some 130 days later than those in the control group. Considering a two-year average life expectancy of rats, 100 days are the equivalent of about ten years in human life.

MSM and Diabetes

The sulfur-containing B vitamin biotin is a critical part of glucokinase, the enzyme involved in the utilization of the sugar glucose. Sulfur is also a component of insulin, the protein hormone secreted by the pancreas that is essential to carbohydrate metabolism. Lack of nutritional sulfur in the diet can result in low production of biologically active insulin. Studies indicate that MSM improves cellular glucose uptake by improving cell permeability, thus balancing blood sugar level and returning the pancreas to normal functioning

MSM RECIPES

MSM + Green Tea Toner for Acne

Ingredients

- 30 ml Aloe Vera
- 28 drops Green Tea Extract
- 1/2 tsp MSM Powder
- 1/4 tsp Vitamin B3 Powder

Instructions

- Add the Green Tea extract, MSM powder, Vitamin B3 and small amount of Aloe Vera to a small mixing bowl and mix together
- Transfer the mixture into a 30ml spray bottle (you may need to use a funnel or dropper)
- Add more Aloe Vera till the 30ml bottle is filled
- Secure the spray top and shake the mixture to blend
- Keep in the fridge and use within 7-10 days

MSM Lotion or Cream

To make 8 ounces of aloe vera mixture with 10% MSM you'll need the follow:

- Kitchen scale
- 7.2 ounces of aloe vera gel (8oz x 90% = 7.2 oz)
- .8 ounces of MSM Organic Sulfur Powder (8oz x 10% = .8oz)
- Blender
- Spatula or mixing utensil
- Plastic baggie
- Air tight bottle or other container

Instructions

- Blend the .8 oz MSM powder (this volume will change depending on the total weight of your end product, and the percentage of MSM you'd like to include - most lotions or creams will allow for 5 - 15% of MSM Organic Sulfur to be added without significantly changing the texture). Alternatively, you can use a mortar and pestle to crush the MSM crystals into a fine powder. Take care not to blend the scoop or desiccant found within each package.

- Mix blended MSM into your aloe vera gel base.

- Scoop the final mixture into a plastic baggie. Cut off one corner and kneed your cream/lotion into the air tight container of your choice. If the container is wide mouthed, you may not need to use a baggie to limit spills.

- Try it out, you now know how to make your own MSM gel. Use whatever cream/lotion base you'd like. Consider coconut oil or shea butter for the main base.

Where Is Your MSM Made?

Where your MSM is made says a lot about it. The great majority of the cheap MSM on the market in North America, Europe, and Australia comes from China. Chinese manufactured MSM will account for much of what you will find on store shelves, where high sticker prices turn away casual buyers who don't research the difference. Chinese-made MSM, however, is generally not distilled but uses a crystallization process that traps contaminants. Either way, it suffers from lower purity levels, a higher presence of unwanted chemicals left from the manufacturing process, and a higher water content, which leads to decomposition and increased contamination. Chinese and other third-world factories undergo few of the health and safety inspections and safeguards ("Good Manufacturing Practice" or GMP) that certified western manufacturers are obliged to follow. They also frequently use lower quality raw materials to save costs.

A consumer who cares about quality MSM should look for a product made in a dedicated FDA or EU certified cGMP (current Good Manufacturing Practice) compliant facility. Choosing an MSM made in a dedicated facility minimizes the risk of the product being contaminated by chemicals, such as benzene, toluene, or pesticides that may be left over from other products that share the same facility. Look for an MSM that has an FDA GRAS (Generally Recognized As Safe) certification. If you maintain kosher or halal, the better brands are generally inspected and certified by the appropriate religious authorities. Responsible manufacturers will also seek third-party testing for their finished products. These safeguards will ensure that each batch made is put through extensive analytics to ensure the absence of lead, arsenic, cadmium, aluminum, mercury, and microbes. This will be topped off with four tests that will establish purity, melting point (important to verify quality), water content (low water content is essential for stability), and freedom from residual DMSO.

Side Effects of MSM

Chances are it is safe if you take up to 6 grams of MSM by mouth for six months or fewer.

There isn't enough information about MSM's safety when you apply it to the skin.

So far, studies have shown minimal side effects when MSM is taken orally in a dosage of 6 grams daily for six months, but some people may experience mild gastrointestinal side effects such as

- Diarrhea
- Abdominal discomfort
- Nausea

Other side effects include:

- Swelling
- Fatigue
- Difficulty concentrating
- Insomnia
- Headache

WARNINGS

Don't take any chances if you are pregnant or breastfeeding. Doctors don't know enough about the safety of MSM in these circumstances. So it's best not to take it. Since MSM is a sulfa drug, do NOT take it if you have allergies to sulfa.

Drug Interactions There are no drug interactions, herbs, supplements, or foods listed for methylsulfonylmethane. The FDA does not regulate supplements. Be sure to tell your doctor about any supplements you're taking, even if they're natural. That way, he or she can check on any potential side effects or interactions with medications, foods, or other herbs and supplements. Your doctor can let you know if the supplement might raise your risks.

CONCLUSION

MSM (methylsulfonylmethane) is an herbal supplement made from an organic sulfur found in foods (grains, milk, vegetables, and fruits), and the urine of animals. Some think MSM is helpful in treating a variety of medical diseases, conditions, and other health problems. MSM and other herbal supplements are not approved by the FDA for the prevention or treatment of medical diseases or conditions. Talk with your doctor or pharmacist before using this product.

Methylsulfonylmethane is available as tablets and powder, and is a component of many herbal products. Methylsulfonylmethane should be stored at room temperature, 20 C to 25 C (68 F to 77 F).

www.ingramcontent.com/pod-product-compliance
Lightning Source LLC
Chambersburg PA
CBHW050252220526
45465CB00002B/647